EXPLODING THE BIG BANG

A BOOK ABOUT THE CREATION-EVOLUTION ISSUE WRITTEN BY A CREATIONIST

JOYCE C. SWANSON

Masterpiece Creations

Copyright © June, 1998
All rights reserved

Masterpiece Creations
27450 481st Ave.
Canton, SD   57013
(605)743-5972

Printed in the United States of America
by Thomson-Shore, Inc.
7300 West Joy Road
Dexter, Michigan   48130

ISBN 0-9664916-3-7

Dedicated to God—
Creator, Maker, Designer

CONTENTS

THANKS

God—Creator, Designer, Maker, for the inspiration for the book and for keeping me motivated to finish it. More importantly, I thank God for life itself, and for introducing me to Himself.

Don & Arla Baker—former science teacher and owners of a Christian bookstore—for reading and analyzing this script, and for introducing me to my husband, who was also writing a book

Dr. Bob Kiner—for evaluating the text for suitability to public school use

Jim & Audrey Behymer—friends from church—for their early encouragement to proceed with this book, and for their "listening ears" when I needed an audience for editing purposes

Jayne Fogarty—my youngest sister—for her many suggestions for titles—her many long distance calls to say she "thought of another one."

Leann Koford—another sister—for her constant encouragement and recognition of the value of this book—and for countless ideas on titles

Vanessa Swenson—a niece—for listening to me read the book to her for editing purposes

Roger Swanson—my husband, whom I met just before finishing this book, for his unconditional love for me, and for his assistance with the final stages of editing

Did boxes evolve? How about pens, chairs, books or dolls? What about people? In a light-hearted, logical way, this book explores the possibility of evolution, or rather the impossibility of it. Using simple, ordinary objects, it demonstrates the absurdity of organized objects accidentally being formed and reproduced. The logical conclusion is that there must be a maker, an organizer, a designer, a creator, God.

First of all, what in the world is meant by evolution? It depends on whom you ask. Absolute evolution says that first there was nothing. Then nothing turned into dust and gases and energy. Next the dust and gases formed non-living things. The non-living things became alive, and lower life forms turned into higher life forms, resulting in people. That gave us the entire earth and everything on it, including all of the ground and seas, all plant and animal life on the ground and in the seas and in the sky, and all people.

To other evolutionists, the world started out from a primitive state of dust and gases, which weren't alive. They don't say how the dust and gases got there. Then those non-living chemicals somehow became alive and became early forms of sea creatures. Those sea creatures turned into land life, which changed into apes, and eventually changed into people.

In this book, evolution refers to the process of life coming from non-life and developing into all the forms of life we have today, including people. Evolution here refers to organization without a director, since the existence of God is denied. If evolution is true, all matter came from non-matter, all life came from non-life, and all higher life forms came from lower life forms. Design in this book refers to the purposeful arrangement of individual parts to accomplish a particular function, requiring a designer.

Some people might wonder why this book gives so many examples of ordinary things that couldn't possibly evolve. Wouldn't just a few examples explain the point very well? That just depends.

For some people, one or two examples would be enough. For a very skeptical person, more examples from a variety of "fields" might be necessary, so examples that cover a wide variety of occupations and interests are given. To an architect, an example of a skyscraper is useful. To an author, an example of a book gets the point across. An auto mechanic will relate best to a story about a car. The more examples in a person's awareness, the better he or she will be able to explain the issue to someone with a different background. So, while some of the sections may seem a little repetitive, it was necessary to carry each chapter somewhat to conclusion, so that any chapter would stand alone, if necessary.

2 *EXPLODING THE BIG BANG*

The main idea of the book is covered in Part One, where the examples are generally progressive in complexity, starting with non-life and ending with life. Part Two gives the logical conclusion to the ideas pondered in Part One, and tells how the author came to be a creationist. Part Three presents more examples to show the extensiveness of proofs in our daily lives. There are examples from every area of life to explain the impossibility of time and chance, of evolution creating our world, with no direction from a Creator. Readers will likely start noticing many more examples in their own world. They will see much evidence of creation.

PART I

THE BASICS

COULD A BOX EVOLVE?

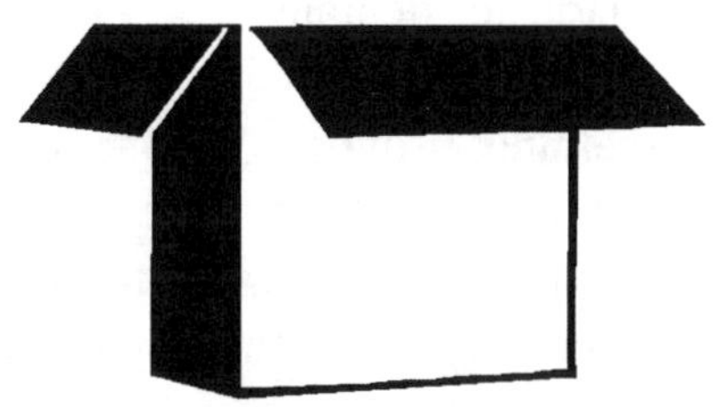

What in the world do boxes have to do with evolution? Plenty, as you'll soon see.

Picture a box. Any simple, cardboard box will do. We're going to study many details of the box, so you will become very aware of it.

In your mind, take a good look at the box. You've likely never really considered what it takes to make a box. You know it has four sides and a bottom, maybe a top. If it didn't, it wouldn't be a box. It may be decorated with pictures or words. It may be empty or it may store something of yours. My question is, where did it *come* from? Oh, you say, that's easy. I bought something and it was packaged in that box. If it's a fancy box, maybe the box itself was a gift. Okay, but how did it come to *be* a box? Well, you suppose someone *made* it into a box. Maybe it happened on an assembly line. Who really cares? I care, and I hope you will too. More than I hope you *care*, I hope you'll see much deeper into what this box tells about how *you* came to be.

Okay, let's look at the box again. If this box could talk (say it was a *voicebox*—no, just kidding!), what would it say about its origin? Would this box say it formed itself? What, you ask? Don't be silly. Boxes can't do that. They have to be made by a person or a machine. That's my point. Things *don't* get made accidentally.

The very existence of something proves someone or something made it. Even if it was made by some*thing*, like a machine, a person had to design that machine, so it was ultimately made or planned by a person. Anything that is organized had to be organized by someone. It doesn't accidentally get organized. Pieces don't accidentally come together and work as a whole. Randomness doesn't produce things that work as a unit. Pieces thrown in a pile land randomly, so randomness is natural. If those pieces make something orderly and functional, we can assume there was a plan to it, and a plan assumes a planner. There's some intelligent plan behind anything organized, including something as simple as a box. Either someone on earth or Someone in heaven plans, makes or creates everything, and it's a cinch whoever made the box was smarter than the box. Anyone who makes something is smarter than the thing made.

Let's discuss this box further. It has four sides, and all four sides of this box are in the shape of rectangles. The bottom is shaped like a rectangle. If it has a top, that is also made of one or more rectangles. Now it's time to let your imagination *really* run wild. We take those rectangular shapes for granted, but do you suppose they all became rectangular by accident? What if one happened to be a circle or a triangle? Would the pieces still form a box? Is it possible that some loose pieces of paper were thrown into the air first, formed themselves into perfectly shaped pieces of cardboard, and then landed on the floor in the right place to form a box? Of course, those perfectly shaped pieces of cardboard had to get folded straight across somehow too, so that, when they landed, they would form a box. The corners had to all be at right angles to each other, with no exceptions. Otherwise, the box would be lopsided, and some part of it wouldn't fit with another part. The pieces had to work, to be functional as a box. They even had to be put together in the right order. If the rectangular parts of the box were laid end to end, they wouldn't form a box. They'd just form a long strip of cardboard that couldn't contain what a box could.

Maybe there's even glue in some part of the box. If there is, how in the world did *that* get there? Where did *it* come from? If it didn't meet up with the cardboard at the right time, it couldn't be used by the box. Also, glue used for cardboard tends to dry. It doesn't stay sticky

forever, so it has to do its job quickly. Was it thrown into the air at the same time as the cardboard? What if it landed on the wrong part of the box? Does that mean the box would end up unglued? The very thought of it makes *me* come unglued! Then again, who threw the cardboard and glue into the air? Oops, that must mean a person was involved, and we said it had to be accidental. Maybe a gust of wind threw them together, but at the same time? Wouldn't that wind more likely cause them *not* to land in the perfect shape required to form a box?

What if these pieces of cardboard have words on them? If something was shipped in the box, there likely are words on at *least* one side. How did the words get there? Were a bunch of letters or some ink thrown into  the air at just the right time, and then did they land on the cardboard in the right places, and in a straight line, to form those meaningful words? What are the chances of that happening?

What if a flood of water happened to land on the cardboard, say from a rainstorm? Would the box hold together? What about an accidental fire? How long would the box last if a lightning storm or tornado hit it and undid it all in a hurry?

In the above example, we start with pieces of paper and then cardboard. What about if we don't even have the pieces of paper? What if we only have some thin pieces of plastic? Oh, that's right. Plastic hasn't evolved yet.

Suppose we have to depend on a gust of wind to bring the pieces of paper to us. Where would enough pieces of paper come from? What if the pieces that come in happen to be a mixture of types of paper? Maybe some pieces are thick and others are thin, If the paper is too thin (maybe only a few pieces of tissue paper), and doesn't happen to get thickened or corrugated into cardboard, it won't hold its shape very well, and likely won't be able to ship anything very heavy. How does the paper become corrugated for the box? All of those dips and ridges are lined up so nicely and make it sturdy. How do those ridges accidentally get the same heights and depths to them? Did the tree accidentally land that way after a billion years of strong winds? Come on now, we have no time limitation. Couldn't it happen eventually? Couldn't it just evolve?

What if the paper that blew together happened to come out of a wastebasket in someone's office? If it's paper, the paper might be a combination of newspaper, scratch pad, stationery, and catalog paper. Would all of these pieces mix together  correctly to form the cardboard for the box? What if this box was formed out in the woods, and pieces of paper had to first be formed out of the trees? Given a few billion years, could the wind or some type of erosion form paper out of the trees, and then blow those papers into the shape of a box? Come to think of it, the tree would have to evolve too, to give us the paper. If

___

everything came from gases and dust, as some evolutionists say, then the trees and wind had to come from that too. All plants and earth and natural laws had to evolve, if there is no creator. It seems strange to even use the expression "natural laws," since laws imply a lawgiver. Even laws are made. It seems that everything is made.

Looking at the box again, if I threw this completed box back into the air and the pieces fell apart again, would they again land in the form of a box? The pieces are all there, so they are available. They should reform correctly again. If the wind tossed them together in the first place, and something held them together in the first place, given enough years, it should happen again.

Could this box have formed itself? Absurd, you say! And of course you're right! These things couldn't possibly happen! But—how about if these pieces were thrown together over and over for a year—a hundred years—a thousand years—a billion years? Wouldn't it *eventually* happen? And then, if it *did* happen, couldn't it happen a whole bunch of times—to make a whole *bunch* of boxes exactly alike? If a box was formed accidentally by pieces of paper being thrown into the air, couldn't it be duplicated by throwing some more pieces into the air? In the randomness of things flying through the air, isn't it *more* likely that things will land in the *wrong* places? It seems there are enormous possibilities for that to happen, and then I won't end up with a box at all.

So, could I get smaller versions of the box the same accidental way I got the first one? Is that where baby boxes come from? Can a box repeat itself—reproduce itself? After all, usually there are a lot of boxes that look alike, especially if the boxes are made to ship something. Given enough time, say a few billion years, wouldn't there be a lot of these boxes around, all the same size and shape? Or, does time have anything to do with it? If something is impossible, can't we just give it *more* time so that it *eventually* becomes possible?

In the world of boxes, could a box ever evolve into a boxcar? The same basic shape could be used, and the material used could be wood, so that part wouldn't even have to evolve into paper first, as the cardboard box did. The boxcar would just need to evolve wheels capable of moving it. If people evolved from something non-living and non-moving, a silly little box ought to be capable of evolving into something that can move. It seems it would be so much easier to evolve a wheel that could move a boxcar than to evolve legs that could move a person.

What if the box started pulsating, coming to life? Now, that would *really* make you think you were going crazy. Is that where baby *humans* come from? Do a lot of baby parts get thrown into the air and accidentally come together and just start breathing and pulsating? Do all of the loose parts flying through the air just get organized? I wonder if evolutionists look at life as just one accident after another waiting to happen.

Is that how this book formed? Did I just throw a bunch of words or letters into the air and hope the air currents would move them around in such a way as to organize the thoughts into sentences and paragraphs? I like to think it's quite organized, since that's the main idea of this book.

I'd say organization proves there's an organizer. Organization of this book proves I organized it. Organization of the universe proves there's a universe organizer, God.

Evolve? We can't even get a box to evolve. What does a box say about evolution? Plenty! It says *it's impossible!*

CHAPTER

2

COULD A PEN EVOLVE?

I wonder how long it would take for a pen to evolve. Let's see. We'd need an ink supply and a container to hold it. It would be handy to have a clicker or a cap on the pen so we wouldn't stain our clothes with it all the time. It would also need to be in a shape and weight that is easy to hold. If it ended up weighing a pound, we wouldn't likely use it too often. If it ended up shaped like a triangle or some other odd shape, it might be difficult to hold. It would also be difficult to hold if it happened to be made of some greasy material. It seems there are so many things to remember in planning a simple pen. Somebody must have put some pretty clever planning into it.

If the pen evolved, I wonder where, in the randomness of things flying through the air, the ink happened to come from. I wonder how the ink happened to get contained in the little holder or cartridge that's inside the pen. When the pen was being formed, the ink needed to arrive before the container got closed up, or it'd never write. The ink needed to be the right consistency too. If it was too thick, it would just glob up and not come through the little hole. The ink had to  have some color if it was to show up on white paper, or else it would have no purpose. I guess if it happened to be white ink, it'd still show up on colored paper. I hope colored paper had been made—oops—evolved by then. I wonder where the color came from to dye the ink or the paper.

It's lucky that the barrel has an opening on the end. It's good that the opening isn't too large, or the ink holder would fall out. It sure is convenient that the ink comes out just a little at a time. That keeps it neater. It's a bonus that it dries quickly too. If it took several hours to 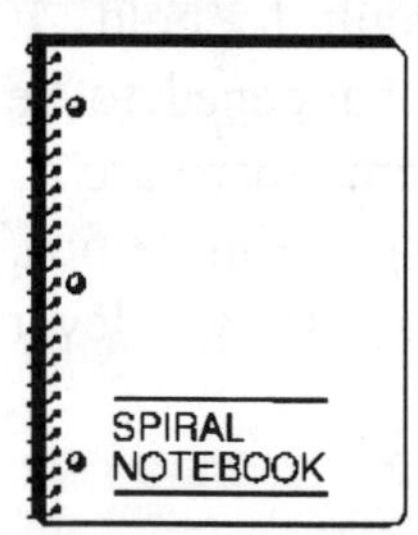

dry, like some kinds of paint, kids wouldn't be able to use it in school very well, where they have to write in several sections of a notebook in one day. I wonder how that quick-drying ink was arranged. That must have just been luck. You sure can't put limits on the wonders that can happen when all of the elements and particles in the universe can arrange themselves in any way they want. Those elements and particles and the wind must have come from somewhere too, but I haven't figured that one out. Unless—there's some kind of mind behind all of this.

But then it wouldn't have evolved out of nothing, and evolution says it ultimately happened that way.

Pens usually have painted cases. It'd be anybody's guess how that paint got onto the case, and so evenly colored yet. Why isn't it blotched? Why didn't part of the case get missed while the paint happened to be going by? Where did that coiled spring come from—the one inside the case? The clicker wouldn't work without it. The spring just happens to be the right length and width to fit the case too. What a marvel that it flew into the case at just the right time, after the pen case was hardened (oops, another thing to arrange!), and before the ink container was loaded inside. Some pens even have ink erasers on the other end. I suppose it takes a different kind of eraser than a pencil eraser does. What are the chances of the right material landing on that end of the pen? What if it landed on the pointer end instead, say, by the point snagging it as it went through the air? That'd be pointless! Get the point? (puns intended!) Useless! So, correct parts of a pen aren't enough. The parts still have to work, to function. They have to be put together in the right order. If the tip of the pen is put near the top of the pen (by the clicker) or if it's installed upside down on the correct end, it won't write. There are *so* many things to remember in designing a pen!

Have you ever heard of a pen getting up and walking away? When we misplace a pen, we often jokingly say that it must have gotten up and walked away. But, what if that were true? What if a pen really *could* walk? Oh, come on, you say. Now I've heard of everything! Oh yeah, what if I also said the pen made a conscious decision to walk? Oh *right*! So now we have a pen with legs and a brain. What next? Well, since you asked, how about if I said the pen cried tears of sadness for having left you like this? So now the pen also has emotions and shows them by crying. How preposterous can we get? A pen isn't alive, and the things I'm talking about make it sound like it's alive. Things that aren't alive don't make decisions to walk around and *certainly* don't have feelings. I couldn't agree more, but evolution says things like that somehow happen. What? How's that? Well, evolution says that inanimate (non-living) things somehow gradually change themselves and become alive. It may take a few million or billion years, but time is no factor. We have all the time in the world. Since our pen has to evolve, I wonder if it might someday evolve into a printer. I mean, ink is used in printers too. That seems like a logical extension, given enough time.

If a pen couldn't evolve and become alive, how could anything else? Our hypothetical pen was given human characteristics. Supposedly it actually became alive at some point and then even became *more* human-like. If humans evolved from something going bang billions of years ago, and then odds and ends flew through the air for a few more billion years, forming organized objects eventually, and eventually pulsating with life, then why couldn't a pen do it? Maybe, given a few more billion years, pens will do that. Remember now, this can't involve a person planning it. The pen has to do it all by itself. It has to form itself and reform itself. It has to somehow grow legs and a brain and tear ducts. Maybe it'll grow by food getting into it. How is it that we can't imagine that happening, but we figure it happened with people, who are far more complicated than that? We have people growing and breathing and eating and reproducing, and we figure all of that happened by chance. Do you notice the strange logic?

Wow, maybe this stuff was really created or something. It looks like something smarter than a pen designed a pen—planned it and made it. Maybe Someone designed the universe—planned it and made it. That sure would make more sense. It'd explain things much better. What's the point? I guess that's the point (pen point!) of this story.

COULD A CHAIR EVOLVE?

Picture a simple chair. It's a gray, metal folding chair, but right now it's unfolded, sitting in front of you. If some pieces of metal were tossed into the air, how many years would it take for those pieces of metal to make this chair? When would all of the loose pieces happen to be the right size and shape to form it? The chair would have to balance so someone could sit on it, so the left and right sides have to be equal in size.

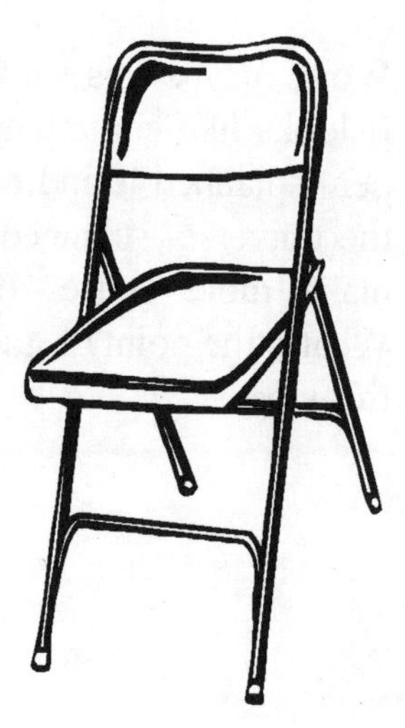

There have to be some screws in the chair to hold it together. Where would these screws come from? Since screws come in various sizes and lengths, how would it get the screws to form out of some random pieces of metal and land in just the right places on the legs to hold the chair together, and to make it possible for the chair to fold up? This scenario is getting pretty screwy, but it gets even screwier.

 I counted the screws in a folding chair today, just to check, and I found nineteen screws in it. There were five on each side of the chair, holding the seat to the legs and holding a couple of braces to the chair. Three screws held the plastic back to the chair. Six screws were under the seat, holding a piece of metal to the plastic seat. This metal is there for sliding the back legs forward when the chair is unfolded. Some of these screws had a Phillips type of top to them, and some were much flatter. I suppose the flatter ones were used to "flatter" the appearance a little. Come to think of it, there were also holes in those chair legs, so there was somewhere to put the screws. And yes, the screws were just the right size for the holes. Oh, I just thought of another thing. There were nuts of some kind on the other end of the screw, to hold it together after it was attached to the chair. Now you know why I said this was pretty screwy.

Let's go back to some of the requirements of the chair. I guess the seat of the chair would have to be attached to the chair somehow, and that seat (again made of random materials flying through the air) would have to perfectly fit the chair it landed on. Maybe that seat even has a cushion on it, and that cushion has some colored material on it. It'd be tricky to arrange those colors to accidentally land on it, especially if there is a pattern to the colors. The chair would have to fit someone who might sit in it, or else it'd have no purpose. It couldn't be too tiny or too huge, or too tall or too short. If the seat were six feet from the floor, a person would have trouble getting up to it to sit on it. If it were only four inches off the floor, it wouldn't do much good as a chair. It would be almost like sitting on the floor, so that would defeat the purpose of it.

A chair wouldn't work if it were made of paper. It wouldn't work if the seat slanted down about 45 degrees to the front, or if it tilted 45 degrees to the left. A person would slide off, again defeating the purpose of a comfortable seat. I noticed there were some tips on the ends of the metal legs too. That must be for protection of the floor. That's a nice feature. It prevents the floor from getting little circles cut into it whenever the chair happens to be used.

Correct parts of a chair aren't enough. The parts still have to work together—to function. They have to be put together in the right order. If the seat of the chair accidentally gets attached to the back support, there'll be no place to sit on it. If the seat gets folded out of shape, say the metal gets severely folded inward, the seat will be too uncomfortable to use. It'd be better for a person to stand than to use that chair. What if a few thin sheets of paper, instead of metal, happened to land near the chair, and that's all the parts the chair had to work with to form the seat? The chair could spend a few thousand years trying to make a chair seat out of paper instead of metal, or it could just hang around awhile longer and wait for more metal to fly through.

There's something else I notice. When I look at a group of chairs set evenly around a table, I wonder how they accidentally got lined up so well. How did they also happen to all be the same size? Did some of  the chairs have baby chairs and then did the mother chair line them all up? Oh, well, nothing is too difficult for a little time to arrange, let's say a few billion years.

Wow! I just thought of another thing! I wonder if this chair could eventually evolve into a wheelchair! It would have the same basic function, only with more parts. It would have to acquire the ability to roll around, again without directions. That shouldn't be too hard, if animals acquired the ability to roll around from just a lot of dust and ashes dashing through the universe at one time. We're not asking the folding chair to come alive or anything. All we're asking is that it accidentally acquire wheels instead of those straight legs. Maybe it could even be motorized, but that's still nothing compared to the "life motor" of animals. Of course, if it *could* come alive, that life motor would have to produce its own energy from some plants or animals that it "eats" and "digests." You know, I'm getting to think that something smarter than a chair made a chair.

CHAPTER
4

COULD A BOOK EVOLVE?

Now, it's time to look at a book. Any book will do. Pull one off your shelf and get a good look at it. How is it made? It has two covers, any number of pages in between, and likely some glue on the spine to hold it together. Let's try to imagine that this book got here through the process of evolution. In that case, the book binding just came out of nowhere and happens to fit this book. The cover is the same size or slightly bigger than the pages. All of these pages "happen" to be cut to the same size, and all of the paper is the same type of paper. Usually it's white. It's a wonder cardboard isn't used for the individual pages, but it probably isn't practical. Oh, but how did that book "know" the difference if it made itself?

Before the paper arrived for the book, it had to come from a tree and be made into a log. Then it had to go through lots of steps to actually become paper. It had to get split "paper thin." I wonder if paper cuts were a problem. Of course, before the paper could even come from a tree, the tree had to evolve, maybe from a weed or a blade of grass. Maybe the book has some artwork, another tricky thing to arrange. Usually the ink used in the text is black, but maybe some of it is colored. Already, the things we've noticed would be awfully hard to happen accidentally—to evolve.

Now, here comes the real tricky part, THE WORDS! It has words on the front and likely a whole lot of words between the covers. Where in the world did *those* come from? What are they? They are just a whole bunch of letters, twenty-six, to be exact,  apparently arranged in some kind of order. Some words are short and others are long. Did a whole bag full of letters get tossed or blown into the air and land in these lengths? How did they happen to form words that mean something to us? Why didn't they land in just a bunch of gobbledygook, in a pile somewhere on the page, and why are they in such neat rows? Shouldn't they be scattered all over the page and some of them be upside down and sideways? Why are there capital letters at the beginnings of the sentences and periods at the ends of most? For that matter, why are there sentences? How is it that these words are in a certain order to form sentences? Usually the noun is before the verb, and every sentence has one of these verbs. Is all of this just luck? How many years did it take for this luck to happen?

Did you say the sentences are even in a particular order—an order that tells a story? How could this be? Could a story just happen because a lot of sentences got thrown together on a page and a whole bunch of pages got thrown together? Doesn't someone have to plan a story? How is it that the words mean something? Does  that mean some creature has learned the language in the book? Do all of the creatures speak the same language? If they speak more than one language, do words from several languages land in the same book? Do all languages use the same letters and have the same number of letters? If the words in this book are all random, there certainly could be many languages represented. It's pretty amazing that the book has chapter titles too. How did those accidentally-placed sentences know what chapter titles would be appropriate, and how did it know to place the titles at the tops of the pages? Why aren't they written down the sides, or upside down, for that matter? Oh, and you say there are other copies of this book? How could all of this happen more than once?

What if this book happens to be a dictionary? Then all of the words are in alphabetical order, and, pray tell, how was alphabetical order determined? There must be half a million of these words and definitions in some of the big, unabridged dictionaries. That must have been some monstrosity to arrange, accidentally, or course.

Could a book evolve into a movie? It'd still be the same story. All it'd have to do is make the characters come to life. If we once evolved from chemicals to animals to people, we should be able to go from a book (evolved paper) to a movie (representing evolved people). We sure take a lot for granted about how books come to be. It's getting to look like it takes something much smarter than a book to make a book.

COULD A DOLL EVOLVE?

Picture a doll. This doll has long hair and is wearing a dress. Other than that, she's not especially remarkable. But wait—isn't she already remarkable? She has two arms and two legs. She has a head and a trunk. Her head has hair, two eyes and ears, a nose and a mouth. She likely has eyelashes. Her arms have hands and fingers and even fingernails. Her legs have feet and toes, and likely toenails. How does it happen that she's so symmetrical? She has the same number of fingers on each hand and toes on each foot. Both legs are the same length. So are the arms. The eyelashes are over the eyes, and not  on some other part of her body, such as her feet. She has long hair on her head. That long hair looks good there, but it sure wouldn't look good on her arms or legs. Most of her body is made of plastic. Wow! The evolution of that plastic must be a whole story in itself!

Let's say the doll's dress is made of cotton. I wonder how the cotton got from the cotton field to this doll. I wonder how it got formed into material, then cut and sewn with thread in all the right places to fit this particular doll. Dresses usually have color and designs in the material, so there'd have to be some dye flying through the air in the evolutionary process of this material being made too. The dress may have some snaps on it to hold it on the doll. The arrival of those snaps would be especially difficult to arrange accidentally. Did I mention that she's wearing a hat and shoes too—just a few more minor details to work out? I wonder how many millions of years it took for this doll and her dress to form from all of the cotton and plastic and metal and dye getting tossed around in the air.

Could she become fully alive? Given enough time, could this doll grow? Can she become an adult doll? Can she have baby dolls? Can she eat real food and repair damaged plastic parts of her body? She has some of the same parts as a person, so why not? If she needs more parts, she can get more parts the same way she got what she has. She can evolve them. Given enough time, could she go from being a little "life-like" to becoming a real, living person? What would stop her? All she has to do is get those non-living parts to acquire life. For that matter, why can't we get a dead person to come back alive? The dead body has all of the parts necessary for life. The parts are even arranged in the right order. All the body needs is life. If it evolved into life in the first place, why can't it do it again? Why does anybody have to stay dead?

It appears that something smarter than a doll designed and made it and its clothes, just as something smarter than a person designed and made people.

COULD A PERSON EVOLVE?

It's time to discuss the masterpiece of design, the most complicated thing created. Let's see if this could have evolved. In order to have life, we'll need the correct material, correct proportions, and correct arrangement of the material, so it's no small order.

So, in a similar vein, (and yes, pun intended) imagine a person. We're no longer talking about a doll, a non-living thing. We're talking about a doll that has come to life. Oops, dolls don't come to life. They may look like the real thing, but they never can be. But, for

evolution to happen, don't non-living things have to become alive? At some point they do. If, in the beginning there were only non-living chemicals that could form and reform endlessly, these chemicals had to form into lots of non-living things. Later, in order to have living things, these non-living things had to acquire life. That means life would have to come from non-life. Is that possible? Scientists call this spontaneous generation, and they say it's impossible. If it's impossible, then people couldn't have evolved from dust and gases, as evolution *says* happened. I've sure never observed a doll coming to life, or for that matter, a pen or book or chair taking on life characteristics, but, for evolution to be true, non-living things had to make regular adaptations toward becoming alive. It would be like a doll growing, replacing cells, and even having baby dolls! The doll would be moving, eating, breathing, reproducing, and reacting to the senses. Boy, this is getting more complicated all the time.

Dolls usually don't have internal organs, but this doll sure would have to evolve a lot of them to "get a life!" Even if the doll managed to get the parts, the parts would still have to work together. The organs aren't just a bunch of spare parts thrown together. They depend on each other to keep the body alive and working at its best. Each has a job to do, and many of them have to be on the job at the same time. For instance, the heart has to work at the same time as the lungs and the digestive system. The nerves have to be in place. A brain has to run it all. The whole nervous system has to evolve at the

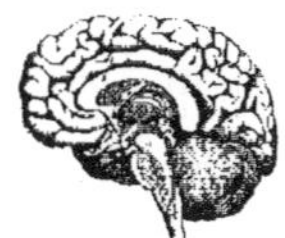

right time. Actually, it probably has to evolve before the rest of the body, since it forms so early in the development of a baby before it's born, and since it constantly checks on everything in the body to see what the body needs. I wonder how the body, and under no one's direction, arranges for all of that to happen.

Since these body systems are interdependent on each other, they need to evolve together. Otherwise, one system will die off while waiting for its partner to evolve. I don't suppose the body has much use for a heart without the rest of the circulatory system or a stomach without more of a digestive system. For that matter, both digestion and circulation are necessary for life, so both systems will have to evolve at the same

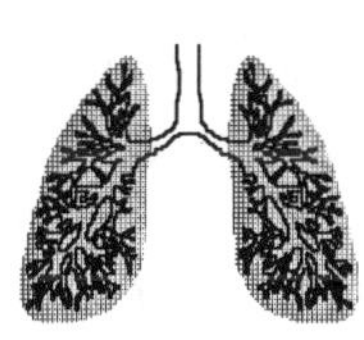

time. The body will need the digestive system to break down food to use. It'll need the circulatory system to send the food, oxygen and waste products through itself. It'll also need a respiratory system to provide oxygen to the body and to get rid of carbon dioxide, and an excretory system to remove wastes from the body. Parts of the body make hormones that are used in other parts, so the endocrine system will be needed to get those to the right places. The reproductive system is needed to make babies so the human race can keep going. If these systems are successfully evolved, the body will need bones to hold itself up and to protect the fragile nerve tissue, and some muscles to move those bones around. It'll need some tendons and ligaments and quite a bit of skin to hold it all together. Of course, all of this has to happen without a plan. No information can be fed into it. Only happenstance can be on the scene.

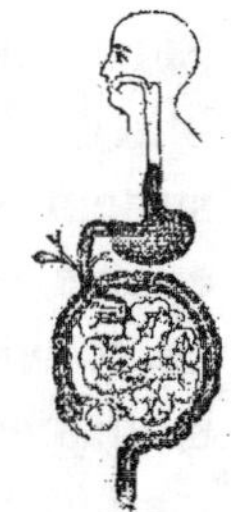

Let's think a little more about the digestive system. Apparently this body sloughs off millions of cells each day, and the body has to make new cells (new parts) to replace the ones that are lost. It does that through the digestive system. People eat meat, grains, vegetables, fruit and dairy products, though the amounts of each of these varies from person to person and from country to country. Some people eat meat and others don't. Bodies everywhere have to use *whatever* people eat to produce the same body parts, such as skin and bone cells, blood, hair, fingernails, eye tissue, energy, babies, and so on. It's incredible that such varied diets produce the same body parts. That must be *some chemical challenge*! If we took pieces of dead chicken or fish or hamburger or fruit or veggies into a lab and tried to make living cells out of *any* combination of those foods, we'd die before we figured out how to do it. Yet the chemical labs of our bodies do that all the time.

In order for the parts to get the food, it has to be delivered from the intestines to the rest of the body.

Along the way, some toxins and dead cells have to be picked up and eliminated. The liver, kidneys, urine, sweat glands and large intestine help with the removal. All of this is coordinated by the brain and nervous system. The nerves keep the brain informed of conditions in the body, and the brain reacts accordingly. It helps if the nervous system is capable of feeling heat, pain, cold and sensation, and if the brain is capable of thought. What I want to know is—how did the brain learn how to do this? Who taught it or designed it to do this? Since *we* can't figure it out, it's a cinch it wasn't one of us who did it. We can't even take a bit of rice or meat or fruit or dust and turn it into a human cell, so how could we design a body?

If I put fish and salad into my body, my body will turn it into body parts and energy. If I put fish and salad into a doll, it will rot. It won't make the doll grow bigger. If I put that same food into a box, the box won't grow. It'll just rot, and it will probably rot the box too. If I try to "reproduce" a box by putting a small piece of paper into a large box, hoping it will grow, I can wait forever, but the paper will never be "born" as a baby box. It's stuck with being the same size it was when it started. Live bodies sure are different from non-living objects.

The reproductive system of living bodies makes completely new bodies, sort of miniatures of the adults. Of course, some of the babies are males and some are females. That must have been particularly difficult to evolve, since some of the parts are different, and since males and females had to exist or evolve at the same time in order to keep the population going. How could people evolve before they had reproductive parts? Why did they evolve as male and female? Wouldn't it have been easier if there was just one sex? Why complicate it with two? Maybe later we'll evolve into three or more, if we can just figure out the advantage to it.

If ears evolved, I wonder how all of the pieces necessary for hearing happened to evolve together. If the inner ear bones didn't evolve at the same time as the outer ear, the person would not hear. If the parts of the eye didn't evolve at the same time that the nerves from the eye to the brain evolved, that system wouldn't work. How did skin sensation happen to evolve? How about the tongue and taste? How about the nose and smell? Who planned these senses and why? It's hard enough to get *parts* to function, but senses? Does that make sense (pun intended)? Sounds like nonsense to me. Do boxes, pens, chairs, books and dolls taste, feel, smell, hear and see? Why not? How did the senses get connected to the brain, where the information would need to be interpreted? How did the brain and nerves evolve at just the right time to accommodate that?

How did bones happen to form around the brain and spinal cord? It appears that the brain and spinal cord are more fragile and need more protection than other parts of the body, so these delicate parts are protected with the very hard substance of bone. Was this just luck? If they'd been covered only with skin, we'd have a lot more people with brain damage and a lot more people paralyzed from minor falls. I suppose, in the process of evolution, that lots of animals started out with something other than bone over their brains or spinal cords. These animals likely didn't live long, since any kind of fall would seriously injure them. I wonder how many types of material their bodies tried out before they discovered bone. That, in itself, could take a few million years. Also, the spine is made up of a whole stack of moveable vertebrae, instead of solid bone. I wonder how many experiments the animals went through in determining how many bony segments were needed for the spine. If the spine were solid, the animal wouldn't be able to bend or twist, so I suppose it had something to do with how much bending and twisting the animal needed to do.

What an incredible masterpiece the human body is! No man-made machine can grow its own parts, find its own fuel, and move from place to place without outside directions. Yet, this living machine does it every day, with everything perfectly organized and timed. This living machine works day and night, every hour of every day, and never takes a day off. Even when it rests, it only rests part of itself. If the person is dreaming, the mind may be very busy even while the body sleeps. While sleeping, the human machine uses fuel that it has refined for itself, and circulates this fuel all over itself to keep every part functioning. While awake or asleep, it constantly replaces itself cell by cell, except for the gray matter of the nervous system. That's the "brains" of the system, telling the rest of the body what to do all of the time. If the body happens to belong to a pregnant woman, this machine is even making another human "machine" similar to itself.

Imagine man inventing a machine that could do all of these things. We'd say that was a pretty intelligent man. Come to think of it, it must be a pretty intelligent God who did all that, and keeps doing it year after year. I wonder when He'll get tired of man's denying His existence. Our questioning of Him seems so foolish, like a painting questioning the existence of an artist, or a box questioning the existence of a designer of the box.

One book I read said that we have fifty trillion cells in our bodies. Another book said we have 100 trillion cells. I can't even picture that large a number of anything, but I know it'd take a lot of years to for that many parts to assemble themselves accidentally in a body and to work as a unit, especially without any direction. Could you imagine fifty trillion people working together without anyone telling them how or when or where or why to get together? They'd all have to have the same goal of keeping that unit together, with a lot of agreement on how to make that work. This would have to happen without anyone organizing the event, and they would have to keep it up for maybe 100 years, if the individual lived that long.

Maybe people are not done evolving. Maybe they will evolve into something else. Who or what says we are done? What will we evolve into next? If we've changed so much in the last million years or so, maybe the next million years will show even more dramatic changes.

Take the eye for instance. We feel certain that two eyes work better than one, so we're glad we have two. But, maybe four would be even better. Maybe future humans (or maybe they'll evolve into something higher than humans!) will get an extra eye or two on the back of their heads. People have often wished they had eyes in the back of their heads. We're supposed to evolve to make our lives better or more efficient, and it would be easier to see what's behind us if we had eyes back there. So, maybe this will become a reality. Who knows? An extra eye wouldn't do any good though, until it was fully formed. One of the goofiest things is that we'd never know when it is finished, unless we'd go by when we "see the light."

Maybe we'd be better off with an extra ear or two, say on the tops of our heads or even near our feet. If we had ears on our feet, we'd probably hear vibrations of the earth better, and hear the footsteps or "handsteps" of dangerous animals better. (Yes, maybe animals would start having hands where they now have feet.) Why not? How do we know it wouldn't be more advantageous to *lose* parts, such as an eye or an ear? Or, maybe those things would be far more advantageous on *top* of our heads, instead of where they are.

According to the theory of evolution, we used to walk on all fours and then "decided" to walk on two legs, so maybe we'll later "decide" to hop on one leg. Maybe that will be more efficient, or maybe we'll decide to spread it over three. Maybe we could do better without fingernails, or maybe we'll develop hooves to replace them. How would we know we wouldn't end up with twenty legs? There are some insects that have lots more than we do. Maybe we'll eventually get them. For now, it looks like we'll have to be satisfied with two, until our bodies are convinced that we'll survive better with more.

Let's say we got two extra legs. We could use two of them at a time and rest the other two. Maybe we could get a lot more done by working only two of them at a time. If we were competition runners, for instance, maybe we could attain faster speed if we raced two of them and kept the others prepared for a good "run at it."

What would a person do while waiting for more legs to evolve? Suppose he has one leg so far. At this point, he has no idea he'll get two. Does he hop on one leg? It looks like both legs would have to evolve at the same time and to the same length, or he'd have trouble using them. What would he do with partially formed legs? What would he do with two and a half legs or three legs? Wouldn't those extra parts get in the way until thy were fully evolved? Till then, they would kind of be a nuisance, something extra to clean and maintain.

Who knows how much we could accomplish if we had an extra set of arms or even just hands. We often say we only have two hands to get things done and therefore are limited. Maybe our bodies will recognize the need to accomplish more, so we'll acquire four hands.

Is it possible the heart could function better with something other than blood, and that (without any help from thinking man, of course, since it has to happen without a planner) our hearts would naturally evolve into using water instead of blood? Water might be easier to come by in an emergency. Maybe we don't really need the heart at all, and we'll evolve into "heartless" beings.

So, what parts are we evolving now? What new parts are on the way? Then again, if everything came from nothing, maybe we'll return to nothing. Maybe that will be the next evolution.

Though I may appear to be digressing, think with me for a bit. Are we the sum of our parts, or is there something more to us? If we managed to evolve all of the parts necessary for life, would we be alive? In other words, is life the sum of the parts, or is there more to life? Does a corpse eat, grow and reproduce? It has all of the parts, but it doesn't do those things. Only living things do them. Living things have information encoded within their cells, telling them how to continue growing and reproducing. Each cell is interrelated with other cells, and most of the cells keep replacing themselves. When one cell gets damaged, other cells come to the rescue to clean up the mess. How could these cells "accidentally" know how to help out?

For comparison purposes, let's discuss a bug or a fly. A fly that has been swatted, or a bug that has landed on a car windshield (He'll never have the guts to do that again!) has all of the right parts necessary for life. It's just that they are not organized correctly anymore. The chemicals are there, but some of them are  no longer in the right order. But that's no problem, right? All we need is a few billion years to get the forces of nature acting on the parts to get them back in the right order and back to life. Given enough time, the pieces will keep interacting with each other until, one day, that little bugger (fly) will fly through the air again. He has a good start with all of his parts. Sure, they're a little smashed, but at least the parts are there. That's a lot more than the fly or bug supposedly had before it evolved, so it should be way ahead of the game, way ahead of its ancestors. Getting the parts to line up right should be the easy part.

Or, let's say the fly died a natural death. It wasn't crushed. It just died of old age. Can it get going again, become alive again? It certainly has all the parts, and they're even all connected. The only reason it's dead is because the parts are no longer working as a unit. They aren't organized like they used to be. Shouldn't it be pretty easy to just add life to it? If life came from non-life (according to evolution), let's add some life to non-life now and give it a new life. While we're at it, let's do that to a few of the human relatives we've lost to death. Why should there even be death, if non-life can evolve into life? Think of the possibilities! Maybe the next step in evolution will be that all dead people will "get life back into those bones" and keep living endlessly on this earth.

It sounds like there has to be some kind of plan, and some kind of planner, some kind of organization to it. Oops! Organizers and planners aren't accidental or a matter of chance, so we can't consider that. Actually, I don't see how we can consider anything *but* an organizer, a planner. Can dead stuff create life? Can disorder evolve into order? Can something without a brain create something with a brain? Can something unintelligent create something with intelligence? All of this suggests a very organized system, so there must be something or Someone organizing it. Since we have the ability to plan and think, and non-living things don't plan or think, I assume that whatever planned or created us must be able to plan and think. It's pretty tough to give others what we don't have ourselves. We have personality, so this creator planner must have personality. Something impersonal couldn't give personality.

We think humans are pretty creative and smart, because we can make a lot of decisions about our surroundings.

We can change the size and design of our homes, and we can modernize with electrical appliances and fancy plumbing systems. Those are creative improvements. They are things we designed or created, evidences of our creativity. We get excited when we finish "making" something we're proud of, such as a book that took several months or years to write, an artistic masterpiece, a house, a hand-sewn dress, an invention, or a gourmet meal. These things give us a feeling of accomplishment, yet we get to start with the raw materials. These accomplishments are minor compared to what God created. He created the real, raw materials used to make any of them. He even created the person *using* the raw materials.

Some people are artistic enough to draw pictures of trees and mountains and waterfalls. Yet God created the *real* trees, the *actual* mountains and waterfalls. We may sculpt an identifiable person. We might invent a camera that takes a picture of a person, and we're impressed with how life-like it looks. He created the *living, breathing, thinking* human being. He made people and animals and plants, and all of the system "works." How great are the accomplishments of the Creator of the universe!

BUILDINGS AND PEOPLE

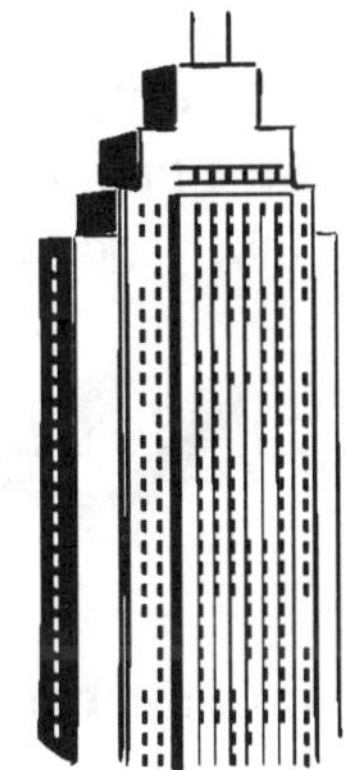

First of all, let's compare the human body to a high-rise building. We would laugh at the impossibility of a high-rise office building existing because of evolution, yet someone has sold us a bill of goods about our bodies doing the same thing. Let's think about some of the comparisons.

Like a building, our bodies have insulation (called fat!). Like a good building, we have a dependable heating and cooling mechanism. Our bodies burn fuel to keep us warm and sweat to keep us cool. Pores in the skin, like windows in a building, help maintain the body temperature. Valves in various parts of our bodies open and close at just the right times to let fluids in. Those valves would be similar to the doors in the office building, controlling the traffic in and out of rooms.

Our bodies have thousands of miles of nerves to keep our brains informed of conditions in the body, so our "telephone" system is state-of-the-art. These nerves are our "lines of communication" between all of the parts. The brain and spinal cord connect or wire all of the parts together. We take this kind of order for granted. But wait a minute. Where did those nerves come from? Were they planned? Who planned them? Do they accidentally connect to other nerves? Who made that happen? According to evolution, those nerves and connections are a matter of time and chance. Could they say the same thing about the phone lines in the building? Would anyone say it was time and chance that produced and connected the phone lines? Who planned them? Do they accidentally connect to other phones? Who made that happen?

What about the water and sewage system in the building? Do we say it got placed by itself? Or, do we say that only the *body's* hydration and waste disposal system got placed by itself?

These are only a few comparisons between a body and a building, yet we can easily see that the body is far more complicated. So how, in our wildest imagination, could we ever say that the body does all of this accidentally, unless we also say that the building got to be this shape and size, and of this complexity, accidentally? Why do we see more need for an architect and builder of a *building* than we do for an architect and maker of a living, breathing, human body? If a living body could evolve, evolution should be a *minor* accomplishment for a building.

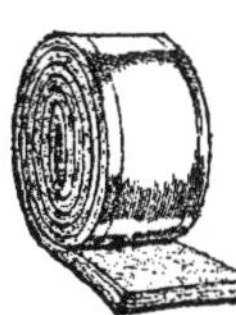

Let's talk more about buildings. If I want to have a house built, I'll need truckloads of lumber, concrete, insulation, glass, lots of nails and paint, some siding, roofing, carpet, quite a bit of wiring for electricity and telephone lines, some sinks and plumbing equipment, some

ductwork, and probably a whole lot of things I couldn't even think of. Someone who does this for a job could come up with a list in a hurry.

But, just having the parts won't make a house. The parts have to be arranged in a particular way. The floors have to be level and the rooms should be squared off with each other. The door needs to perfectly fit the openings left for the doors. It'd be nice if the paint would end up on the walls and not on the floors. Preferably, the electrical outlets would be near the floor, rather than near the ceiling. It'd be nice if the plumbing ended up in the bathrooms and kitchen, mostly. Come to think of it, it'd be nice to have carpet, and the carpet would be appreciated more on the floor than on the

ceiling. That all sounds like reasonable expectations, but would it still be reasonable if it had to happen as the result of time and chance, without planning?

Even a ladder, used in building, would be difficult to evolve. It can't be luck that caused the rungs to be of equal length and of equal distance from each other. It isn't luck that each rung of the ladder is attached to the sides of the ladder to make it trustworthy to climb.

Couldn't I just buy all of the supplies and wait for a windstorm to come along and arrange it into a house? I could set a lot of concrete, wood, glass and wire on the land let the storm toss it around in the air. If I wanted to add a level later, I could toss more wood and such toward the building. Maybe a good tornado would do the job for me. It sure would be cheaper than hiring a builder. I hope it turns out to be a helpful tornado, or else it'll just be a waste of time. There's also the chance that the tornado will wreck the process or deliver it into the next county, and then I'll have to start all over.

I wonder how many years it will take for my dream house to get built this way. I'm looking forward to furnishing it on the inside too, so I hope it's not too long. I want to invite people to see the finished product, and I know they won't want to wait forever. I hope it happens in my lifetime. My friends will probably like it so much that they want one just like it, and who knows, given enough time and enough winds, they just might get a "baby" house, just like mine.

Let's say I was really happy about how this house turned out, and I wanted to have a skyscraper built in the same way. I guess, if a box could evolve, then that box could get *really* creative, I mean "evolve," and end up as a room in a skyscraper. Maybe that's how the Empire State Building came to be.

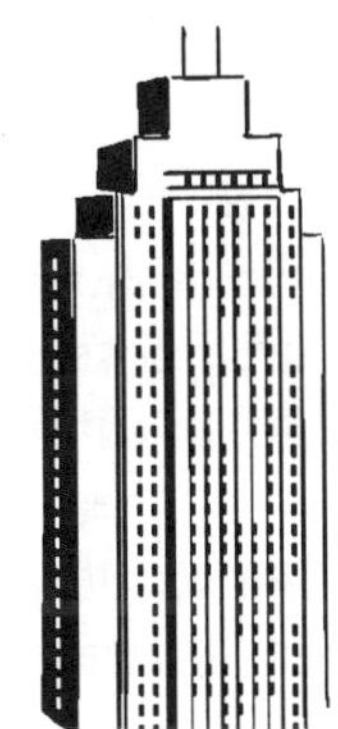

For our skyscraper, we'll start with evolved cardboard boxes. The boxes will eventually become very sturdy and have to evolve into wood boxes. Eventually, some steel girders and concrete (evolved from the original glue in the box) will meet up with the wood and reinforce the boxes where necessary. The building will be heated and air-conditioned from a central location, and duct work will connect the furnace to all of the offices. Somehow, a furnace and air conditioner need to evolve and land at the end of the ductwork. It'll be good if the furnace and air-conditioner don't both try to work at the same time, or the place will never get heated *or* cooled. There will be lots of windows in the building, allowing for individual offices to regulate the temperature that way too. For those people whose body temperature don't match everyone else's in the building, that's

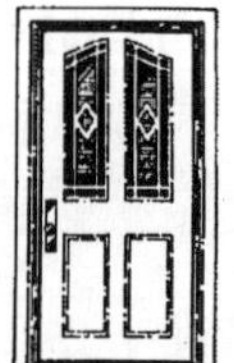

a nice feature. I imagine the owners of the building will be happy that the building manages to evolve some insulation in the walls, or the energy bill would be a lot higher. It'll be quite convenient if there are lots of doors, halls, stairways and elevators, to maintain efficient traffic control. People are happier when they don't have to leave by way of a fire escape, but even that will do in a pinch.

Remember, everything in the building has to form and arrange itself without a planner, and we have all of the time in the world, millions or billions of years, if necessary. We're in no hurry. It just has to happen *sometime*. Remember also, that this building should have an *easy* time evolving, since it doesn't *even* possess life. We only have to go from nothing at all to an entire, furnished office building. Maybe we could even cheat a little and say we have all of the chemical elements necessary for the parts, but they're just scattered throughout the universe. Have fun, randomness!

Just suppose, after the skyscraper finally evolved after billions of years, that an explosion or a severe tornado whipped through the area and dumped all of the parts on top of each other. How long would it take to rebuild, naturally, of course? How long would it take before all of the parts would once again produce the Empire State Building? During that time, of course, there may be more explosions or wind storms, so it might have a few rough beginnings. As it's forming itself, would it continue to develop through those catastrophes? What would hold it together? Maybe it would decay or rot while it waits for the right wind. That would make it worthless. Maybe the steel would rust before the sections accidentally landed in the right places to get welded. It would take a miracle, and miracles aren't natural. We need God for that, and we don't have Him in this scenario.

No one would expect to find order from such storm rubble. It's not likely that anything would be working better after a tornado or windstorm than they did before. We would not expect that piles of bricks and lumber tossed around in a tornado would automatically form themselves into a house afterward. It's not common to get order from piles of stuff tossed around during a wind storm at a construction site.

But, if this could happen, maybe that's just what the world needs. We just need a good, strong tornado or an explosion *to* put some order into the world. We could mix up everything in the world that doesn't work, and maybe we'd get some real good, working machines and new forms of life from it.

DIFFERENCES: ANIMALS AND PEOPLE

Some people have trouble understanding where all the human races came from, since we have different skin colors and some distinctive body features. Those differences are minor, compared to the differences between apes and humans. If we have trouble understanding how all people in the world could descend from a common color, how in the world could we understand or believe that people evolved from apes? Worse yet, how could we understand or believe that we evolved from nothing or from molecules to a sea creature to an ape to a human?

Some animals hibernate, but not all. How would hibernation evolve? How would some animals know it to be advantageous? Did lots of them (bears, for instance) die before they figured it out? Why don't people hibernate?

Our food habits seem to be different from that of the animals. I don't know of any animal that cooks its food, like we do. Also, when animals kill and eat each other, they eat a lot more of the animal than we would. Most of them eat the intestines of whatever they kill, but most people don't.

We think we're pretty intelligent because we are able to design a box or a pen. It's organized and it functions. A bird can't do that. A bird makes a nest, but he has to do it by instinct. He can't design it any way he wants. He can't decide to make it with bricks instead of sticks. He can't decide he wants a high-rise so all his relatives can live with him. He can't set up a phone system in it so he can talk to his distant relatives in Russia. Different birds make different kinds of nests, but each species has to "stick" to his kind of sticks. One bird doesn't build a skyscraper and another of his species put in a basement or decide to color his twigs blue or red.

Can an ant ever make a home of cement or glass? Can it decide to paint its home a new color? A female ape can't decide to wear a different outfit for every day of the week or to cook up interesting meals for her family. Animals go by instinct, rather than creativity and instinct, rather than creativity and decision.

Even if body parts could evolve, how would intelligence evolve, or even instinct? Where did we get morals and a sense of right and wrong? Do animals have morals? Do animals make plans to better the world, to invent things? Did we get our morals from apes? (Let's hope not.) Where do we get our convictions? Is an ape capable of convictions? How did we learn to be unselfish, if our very survival of the fittest depended on us being selfish? Where did we learn how to be rational? From where did our concern for life after death come?

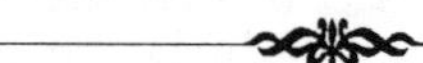

People are able to think and imagine and change the course of their futures. They have the ability to evaluate, deduct, reason, organize, analyze, and make judgments. We can use logic. We are able to decide between several courses of action. We can observe and think and choose an unselfish course of action. We can write poetry, build skyscrapers, build and pilot planes, and program computers. How did we evolve those abilities from apes which can't do those things? How did we evolve from molecules that are further yet away from those traits? How does a thinking, moral person with an IQ of 140, and an author of technical books, come from an animal that doesn't think or have morals or write books? If these animals are in an evolutionary process toward becoming human, they should have *some* of these human characteristics, especially if we came from them.

There are no primitive people today who grunt for their language, so we don't even have a "missing link" in the area of language. Even "backwoods" people are able to express themselves very well, no matter what language they use. It appears all languages are fully developed, that there are no half-way languages.

Even our tongues are special, in the way they are used for speaking. Do animals have any similar use for tongues, or is that one of the ways we are different from animals? We can't think without words—without language. Animals don't speak English or read books. If apes evolved into humans, there should be some apes in a transitional phase, catching up with humans like other apes did, and just about ready to start talking to us in our language. There must be more than physical characteristics that evolved, so why aren't they able to talk to us, say, with a few thousand words, since we have many thousands? Animals are limited to living by instinct, so they don't have all of these abilities. Their instincts serve them very well, but they don't appear to be anywhere near becoming human, and gaining our additional abilities.

If everything is fully adapted (evolved), including animals and plants, how did all of this happen to end at the same time? Maybe it hasn't. Maybe apes or humans will someday fly with evolved wings. Maybe birds will evolve into something completely unknown at this time. Maybe trees evolved from grass, and will someday look completely different, as different from present day trees as our trees look from grass. Maybe they'll someday be "sky-high." Maybe trees will someday start walking or flying or they'll turn into grass. Maybe people will turn back into apes. Then maybe people will become extinct, like the dinosaurs.

How can "chance" produce something beneficial. "Chance would have to know what is meant by beneficial. It would have to have a sense of right and wrong, or good and bad, to produce something good, or fit, for survival of the fittest. Maybe it made a lot of accidents first, things that weren't good or fit. Then where are all of the mistakes? Where are the fossils?

We often speak of looking for these missing links in the fossil record, but we should look for them in the living world too. If there were in-betweens before, there should be in-betweens living now. Apes should still be evolving into people. Why would they quit? Surely all of them didn't complete the process toward becoming human at the same time. Where are all of the present-day, in-between models? Why don't we still have a lot of the in-between models for other animals? They wouldn't likely have all died off naturally. Maybe we could buy the theory that some of them wouldn't make it, but all of them? It seems like we'd have living examples of all the steps, maybe living in different parts of the world, if they happen to need certain climates.

Most of us grew up with a fairy tale about a frog turning into a prince. We readily recognize it as a fairy tale, but to an evolutionist, it could be true. It just took millions of years, but we have the same ancestors as the frog does. In fact, all creatures have the same ancestors, according to evolutionists, so we might just as well say a snake or an alligator turns into a prince.

If evolution is true, then that means plants evolved too. We always talk about animals evolving, but these animals had to eat something. Before there were animals, there would have to be food for them. We don't hear much about plant evolution, but it's a cinch the animals had to eat something to live. Whatever they ate had to have its own evolution. Even the earth, dirt, rocks, trees, sky and air had to evolve. The sun and water had to evolve. All chemical elements had to evolve.

LOGIC IMPLIES CREATION

Maybe you're thinking, "Okay, okay, I get it. Boxes and pens and dolls and such can't evolve, but I never thought they did. Actually, I never gave any thought to how they *did* get here. But people and animals are different. They are alive, and I only thought *living* things evolved."

This section is especially for you. First of all, I don't fault you for thinking this way. Many people have just never stopped to think about how impossible it would be for organized things to get formed accidentally. I'm trying to help you realize that organized things can't be formed without design, and logic implies that anything designed had a designer.

When you say that you thought only *living* things evolved, you're saying you think it'd be easier for a thinking, breathing, moving, growing cell to come together by chance than for a box to form by chance. You are thinking that the *simplest* objects can't evolve, yet the most complicated or complex things *did*. A non-living box with only a *few* parts can't evolve, but a living body with billions of cells did. Chance didn't make a box or a book. It only made the most complicated thing in the world—you!

Does this make sense? Animals and people are far more complex than boxes or pens, which are just *objects* designed by people. They are so minor as to be ridiculous in comparison. And that's my point (pen point, ha ha!). I am trying to show the absurdity of it, by imagining their evolution as a joke.

Logic alone shows the impossibility of people being here as a result of evolution. Logic tells me that boxes and pens didn't come together by chance, didn't evolve. You already know that, but do you see the implication? Something so simple as a box or a pen can't get organized accidentally, can't evolve, can't happen by chance, can't accidentally get organized enough to become functional. So, how in the world, with any stretch of the imagination, could we expect something far more complex and intricate, like live animals and people with billions of parts, to get organized and function accidentally, as a result of chance? If an uncomplicated thing can't evolve, how could something extremely complicated evolve?

Complex things that work as a unit can't be accidental. Adding life to their parts, accidentally, is even less likely, like totally impossible. We can't even get plain old, *non*-living boxes or dolls to evolve over endless time. What makes us think a person with billions of *living* parts (especially on a microscopic level!)would be easier to evolve? These parts are even interdependent on each other. How can we attach a greater possibility of a live thing evolving than we can to a box? Can a box grow, reproduce or move around? A doll has some of the parts of a person, but can a doll grow, reproduce or move around? Can life come together (evolve) more easily than non-life can? Are living, extremely complicated things more likely to evolve through randomness, than non-living, uncomplicated things? It makes no difference if we have a trillion years or endless time. Is it *ever* possible?

The odds against this kind of thing happening have to be absolutely mind-boggling. The idea of an animal or a person evolving should be preposterous, simply illogical. It seems that belief in evolution requires far more faith than does belief in creation. Belief in evolution requires a staggering amount of faith in chance, rather than use of logic. It seems we have to just toss logic out the window to believe we are here because of evolution.

For the world to have evolved, nothing had to make something, and then it had to make more complicated things. For awhile, all of these things had to be non-living and impersonal. Then those things had to become alive. Then those live things had to keep getting more complex, turning cells into tissues and organizing tissues into organs. All of those organs had to work together to keep a particular body alive. Lots of these had to be made with exactly the same parts, so the species wouldn't die off before it had a chance to evolve (or to allow for those that would die before they successfully evolved). Of course, the male and female versions had to develop at the same time, or there wouldn't have been any reproduction. Since those male and female versions had some parts different from each other, that would have been a little trickier to arrange. (I wonder how bodies, without any intelligence guiding them, would know they needed male and female parts to keep this thing going. Come to think of it, I wonder how intelligence would have evolved, since *it* was needed eventually in this process.)

In the normal world, things fall apart. They decay. They degenerate. They don't "build up" without the help of a separate builder. They don't become more intricate on their own. If there's an explosion in a junkyard, a "big bang," we find a bigger mess than we had before. We don't find organized, working objects that were made during the explosion, and we especially don't find things that came to life as a result of it.

Logic tells me that anything organized or made or produced or designed had an organizer or maker or producer or designer. Nothing gets organized by itself—not a box, not a pen, not a chair, not a book, not a doll, and certainly not a person. These things aren't made by chance. If we require a designer to make a box or a pen or a chair, we must need a designer to make animals or people or any other kind of life.

Can *anything* make itself? Can a picture make itself? It needs an artist, and the existence of the picture proves there is an artist. Can a box design or create itself? It needs a designer or maker, and the existence of the box proves it had a maker. Can a book exist without an author? Of course not. Can a person design or create or make himself? The existence of a person is much stronger proof of a maker. If even the simplest things prove a maker, the most complicated things prove an *extremely intelligent* maker. How difficult it would be to imagine a universe without a universe maker.

If the box could think, maybe it would probably think it made itself. It sure wouldn't want any "person" taking credit for the box's success, just like we don't want a creator to take credit for any of *our* success. In the evolutionist world of thinking, without a creator, there can't be anything smarter than the object telling it how to form. In the case of humans, no credit can be given to anyone over us. We have to think we did it ourselves.

Put very basically, there's a whole lot of stuff around, and dead stuff doesn't create life. It had to be made, created, designed. Disorder doesn't evolve into order. Something without a brain doesn't create a brain. Something unintelligent doesn't create something with intelligence. Put a little more eloquently, there is an incredible amount of structure and order and function and design in the universe, so there has to be a creator, a planner behind that structure and order and function and design. Design implies that parts are arranged according to a purpose. That requires a designer. That also requires matter, and matter can't create itself. There can't be creation without a creator. There is no other logical conclusion.

Either a box evolved or someone made it.
Either a pen evolved or someone made it.
Either a chair evolved or someone made it.
Either a book evolved or someone made it.
Either a doll evolved or someone made it.
Either a person evolved or Someone made it.
Either the universe evolved or Someone made it.

CHAPTER
10

MY AWAKENING TO CREATIONISM

I was raised in a family that strongly believed in God, but I let go of any interest in Him for a few years. It wasn't my parents' fault. They assumed I was strong in my belief in God, but I had memorized answers about Him. The answers didn't come from the depths of my heart, as my convictions do now.

My first questioning of God's existence came when I went to college and met people who never gave God a thought. That sounded appealing, like an easy way out of feeling guilty for wrong-doing. If there's no God, there's no accountability for sin. I wasn't sure God existed, and I didn't really care to learn for sure (I thank God for being merciful to me during those years.).

Then I heard a talk about organization in the universe, and I had to rethink things. I became absolutely convinced there was a God. I saw that as the only logical explanation of so much organization in the world. I wondered why no one had explained it to me this way before. Now it seems only logical that I explain it the same way, and that I apply it to the issue of creation verses evolution.

I can understand a little randomness in the world, but not enough to explain the countless examples of things being organized. If little things like cardboard boxes and pens and chairs and books and dolls couldn't form themselves and evolve, how in the world could people and animals evolve? If it takes people to design simple, inanimate things, how can complicated things like plants and animals and people be designed by "nobody"? These "accidental" things are too amazing to be mere chance.

I figure I can save other people a lot of time by making this concept very easy to grasp. I sure would have appreciated someone doing that for me earlier than they did. I don't think a person has to be a scientist to understand that the earth was created. It's just a matter of logic. In fact, I don't see how a logical person can believe it evolved. Can life come from non-life? Do things accidentally get organized and then reproduce themselves in this same accidental manner? Do ordinary objects like boxes and pens grow larger because food is put into them? If such simple objects require a designer, wouldn't much more complicated things like people who live and breathe and digest and reproduce require a designer? I believe God must be so incredibly amazing to figure out how to do all of this. For this reason, I dedicate this book to God, my Creator, Maker and Designer.

PART III

ADDITIONAL EXAMPLES

A CAR FROM A JUNKYARD

If a tornado moves through an auto junkyard, could we get a working car out of it? If it's a big junkyard, there ought to be enough parts to make one car. The tornado would shake stuff up pretty well. Can it do it enough to build a car? All we have to do is get all of the parts together in one place, and each car part in the right place on the car. The engine has to be connected to the fuel line, and the tires and rims have to get attached perpendicular to the lower four corners of the car. Maybe we'll need more than one tornado. That's okay. We have enough time (as long as we aren't in a hurry to get someplace!). With a few good tornadoes, and whatever other natural disasters we happen to get, there ought to be a working car there someday.

Of course, there is always the possibility that, in the randomness of things flying through the air, the necessary parts will fail to arrive. Parts will end up in the wrong places, unable to function any useful way. Even if all of the parts arrive and get put together (by accident, of course), it still might not work. I just hope mankind is still using the same kind of fuel when it finally gets put together by itself, or else the gas tank won't be useful. It'll probably need oil too, so that will have to evolve and land in the right places of the vehicle. It'll need transmission fluid. It'll need water or anti-freeze, and that'll have to land in the radiator. I'm no expert on cars, but I bet it'll need to be lubricated a few places too. I hope rust doesn't render any of the parts useless before the time is up. Even when we have all of the parts, we still need something to give it "life." So, I guess it needs a spark from the battery, and oh yes, a battery and spark plug. After all of these parts get assembled, I sure hope the correct key accidentally finds its way to the car, or we'll still have trouble going anywhere with it. We'd better have some brakes for this thing. We'd sure hate to not be able to stop it, and have an accident with it right after it got going. Otherwise, we'd have to go through a lot of these steps again to get a working car. It'll probably require some kind of regular maintenance too.

Now, don't forget. All of this has to happen without a set or directions from anyone, just lots and lots of chances, lots and lots of coincidences. We have no plan because there's no planner. We sure don't want to put any limitations on when this gets done. We have all of the time in the world.

We're seeing how hard it is to accidentally assemble a car when we have all of the parts. How much harder it would be if we first had to make each part! Oops! I mean how much harder it would be if all of the parts had to first evolve! We can't make them ourselves, because then we'd be a "maker." For evolution to happen, there can be no maker. I can't believe how hard this is getting! All I want is a simple car!

It's funny about giving the car "life." I use that term because I am suggesting the car can move without me pushing it. When we talk about ourselves having life, we sure expect more of ourselves than that. We expect to grow from little babies into adults by putting in food, but we don't expect the car to start out tiny inside another car, and grow into a full-sized car, just by putting in enough gas. We expect to reproduce ourselves sexually, but we sure don't expect anything like that of a car. We expect to be able to bend and twist ourselves to sit and stand and lie down, but our cars would be wrecked if we bent and twisted and laid them down. We use our brains to plan our days, but we don't expect a car to have a mind of its own and decide where to take us. Some cars have computers in them that might make them *seem* intelligent, but a person still programs the computer. I guess our bodies must be a lot more complicated than car bodies. We look at a car and the many parts to the engine and fuel system and brake system, and we know someone designed the car. We would think someone was crazy if he said these parts came together by themselves. So, if a car can't evolve, how in the world could we?

ROADS AND RAILROADS

I was in Ireland one summer, traveling by bus. I noticed that the land on one side of the road was developed. There were rock fences, rock houses, and rock farm buildings. On the other side of the road was undeveloped land. There were lots of rocks there too, but they were scattered all over the fields. There was no order to *those* rocks. Apparently, something happened to make one side of the road look different from the other side. Apparently, people worked the rocks on the developed side. But, how do we know that the rocks on the developed side didn't just land naturally as fences and buildings? Could a series of earthquakes have formed these fences and buildings? How about if we had a few million or billion years? Couldn't time and chance have formed them? And if time and chance formed them, why didn't it form them on the other side of the road? Why didn't the rocky side make its own fences? Didn't it have "millions of years" to do it? I imagine there was equal time and chance for both sides of the road. Why was one side completed and the other side left natural? One side must have had a designer.

See, we know that we have to plan to get order out of rocks. We know that people designed the buildings and fences. We can see what happened (or didn't happen) on the other side where people didn't arrange the rocks. We see developed land and know someone developed it. We should be able to look at anything organized like that and know there's an organizer. We should be able to look at our organized universe and know there's a universe organizer, God.

I also rode the railways in Europe this same summer. Does the track just happen to meet in train stations in each town? Do the tracks just happen to be parallel? Not only that, but there's exactly the same distance between the tracks within whole countries, except for tracks owned by some regional railroads. The railway people say that's so all of the trains can use the same tracks! If the tracks were a couple of inches closer together in spots, there would be a lot of derailments. Ingenious! What luck that it happened without planning! Time and chance are so smart!

Later that year, I drove home from a creation seminar. I thought about the road I was on. I wondered how long it took that road to evolve into its route. How did it happen that this road led where I needed to go? Here I was, driving from Marshall, Minnesota to Sioux Falls, SD, and there happened to be a road connecting the two towns. How many years did it take for this road to make itself? It must have taken a lot of concrete. I wonder how long it took for the concrete to happen to land there. It's a good thing the concrete landed in a wide enough strip for at least the width of two cars. Otherwise, there could be lots of accidents on that road. It just happened that there were lines down the center of the road too. Those lines happen to be parallel to both edges of the road. On some roads, there were also lines for left turn lanes. I find it incredible that all of these lines were so well-spaced by accident! I bet that *prevents* a lot of accidents! At night, they're especially helpful, and even more so in a snowstorm. How could time and chance "know" that these lines should be evenly spaced across the road? It's a good thing the surface was smooth too. I'm glad there weren't huge bumps and dips in it, say, patches of concrete a couple of feet high or holes a couple of feet deep. I noticed that there were roads connecting other towns on the way too. I wonder how the road happens to connect all of these towns. Since it all has to be a matter of chance and endless years, it doesn't seem likely that towns and roads would be so well connected.

It seems preposterous to consider that a highway could have evolved itself into being, yet some people think *people* did. They imagine that the human body, with its lengthy system of nerve "roads" and blood "roads," is here because of lots of body parts accidentally lining themselves up to a workable unit, yet they couldn't imagine a simple road accidentally lining itself up to a workable unit connecting lots of towns. How impossible it would be for even the main parts of the digestive system (the esophagus, stomach, and intestines), to form accidentally, yet the digestive system is far more complicated, involving every part of the blood system. While we'd be amazed that roads could connect towns, without anyone planning the highway system, we expect the much more complicated routes in our bodies to have organized without a planner. How illogical we can be!

Speaking further of roads, could a map of roads evolve? Maps show what has already happened, but could the map change without someone making the changes? Even a map needs a map maker. No one suspects that the map made itself, that lines happened to line up on the right places to represent every street in New York City, for instance. Did all of the buildings in New York City get there by chance? Did they evolve? There are many more cells, with many more parts, in a human body than there are buildings in New York City. Could New York City need a planner, but our bodies *not* need a planner? Which is more complex? Apparently our bodies are designed according to a complex plan, so there had to be a body planner.

FILE CABINET

I have a file cabinet near me. As I open a drawer, I find maybe fifty files in alphabetical order. I have organized these files so that I can find information I need in a hurry. If I tossed all of the papers in these files in a pile, I'd have a very difficult time finding information I needed. If I left it to chance, maybe these files would all get replaced in their original file folders, in alphabetical order, without me or anyone else organizing them. These files are organized to serve a purpose. If someone wanted to be really nasty to me, he could dump these drawers and throw the papers to the wind. I'd probably never recover all of them. It'd be like throwing feathers to the wind. But then, if evolution is true, and random things organize themselves into useful functions, eventually it'd all be organized again. I may have to wait a few billion years for it to happen. Why is it that we accept evolution for really complicated things like a human body, but can't accept it for our files to all organize themselves?

COMPUTER

Let's say you take a trip to Mars. You land on Mars and find a desk with a computer on it. You immediately assume you are not alone. When you 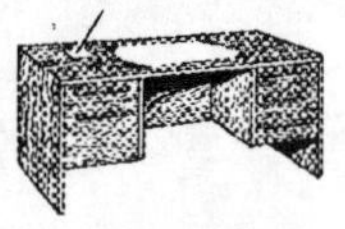 see design, you expect to see a designer, especially if  you see that all of these things are arranged in some kind of order. Even the existence of these things proves order, but if they're also apparently arranged to do work, you'll assume even more of a designer. Let's say there's a chair next to the desk, and a phone is on the desk. It appears that someone intends to do some serious work here, and expects to contact someone else by phone while here. *Unless* it all just happened to arrive here by accident and the phone line *happened* to be connected to phones all over the universe by accident. You'll probably want to check for dial tone, just to be sure it's connected.

We give computers credit for being able to do many things, but people make computers. The computer can't make itself or program itself. The computer can't even decide to take a break. It has no feelings, no emotions. It can't decide whether or not it wants to continue being a computer. It can't build itself into a smarter or faster version. People have to do that. It's only a puppet. As long as the computer was designed by someone, that someone has control over it, and even decides whether or not it exists. The designer makes the rules.

Maybe the computer thinks it made itself, or maybe it thinks it evolved from a typewriter by making new and more complicated parts for itself. Maybe that computer has aspirations of someday making itself into a satellite and orbiting the earth. Of course, it has to do this all by itself. It can't have anything smarter than it giving it orders. The computer probably thinks it's the smartest thing there is, so it's at the top of the intelligence level, and therefore capable of becoming God. This computer has unrealistic ambitions. Whoever made the computer is surely is in charge of it. In the case of the universe, Whoever made it (the Creator) has complete rights over everything in it.

PAINTINGS, DRAWINGS AND SCULPTURES

If we see a framed picture showing land, trees, a waterfall landing in a lake, a mountain and a path, we might assume that someone drew the picture based on an actual scene. I have a picture like that in my home, and I am looking at it as I write this. You'd have a hard time convincing me no one painted it, especially since I watched the artist do it. I wonder which would be harder to do—to paint the picture or to actually create the live trees, the running water, and the mountain shown in the picture. Likely there are some fish in that lake too, but I can't see them. Would they be harder to paint or to actually produce? I never asked the artist to produce the live trees or fish, but I don't think he would have discussed the assignment with me very long. Then again, maybe, given enough millions or billions of years, his drawing will automatically evolve into the actual scenery. I wonder how the artist would have felt if, after he finished the painting, I told him I believed the painting just evolved, that he didn't really paint it. I suppose he'd have been flattered with my keen understanding of how things are made.

I wonder whether it's easier to draw flowers or to make live flowers. It even takes a designer to make or arrange silk flowers, so I suppose the real things require a designer too. I've taken many videos of people, and it's sure easier to record 

those memories on tape than it would be to create the people in them. When I take a photograph, I have only a representation of what a camera can capture, but I sure didn't create the people. A camera is a pretty wonderful thing (again designed by someone), but the human eye that notices the people to photograph is far more marvelous.

So—

- People arrange silk flowers, but God makes the real flowers.
- People produce videos, but God causes the real event to capture on "reels."
- People draw pictures of people and scenery, but God creates the real people and scenery.
- People photograph other people, but God creates the real people.
- People invent tape recorders, but God creates the person talking on the tape.
- People act in movies, but God acts to produce the bodies of the actors.

All of these people are creative in their work of arranging, producing, drawing, photographing, inventing and acting. We don't pretend that their finished products are a result of time and chance. We don't pretend that their work would have happened if they didn't get involved in their projects. Yet, some people think the majestic world of people and scenery and virtues is the result of time and chance.

Historians have found cave drawings of animals. They assume people drew these. No one would say that natural erosion in the caves caused these pictures. There would be too much definition, too much likeness to animals in existence.

That's similar to the sculptures of Mount Rushmore. No one assumes these were shaped by wind and erosion. Not even a billion years could make it happen. Their organization and resemblance to real people, even identifiable people, suggest that someone planned the scene. Now, if the dead people represented on Mount Rushmore stood up and walked away from that mountain, we'd really have a scary scene. Stone sculptures don't get up and walk away. We'd have gone from non-life to life! We can't imagine a transition like that! Or can we? For evolution to happen, we had to go from non-life to life! So, why couldn't it happen again? Why can't our Mount Rushmore sculptures get up and walk away? Could the cave drawings come to life? How about sculptures? If these drawings and sculptures represent people who evolved, didn't those people have to have all their parts first and then get life? What if all of the parts were sculpted into the stone? Would that make them come alive? Even the stony representations had designers. Then how is it possible that the people they represent could have no makers or designers? How could animals and people evolve if drawings and sculptures of them couldn't evolve?

If we see a statue, do we say it got here through evolution? It happens to look like a person. So, do we say that gases and dust formed it and wind, rain, and weather eroded it until it looks like a person? Maybe it even represents someone we know. That must *really* make it accidental! Do we assume that it became a statue over millions of years? Then we'll probably expect it to walk around one day too. We'll expect it to have babies. It'll tell its babies that they made themselves. It'll say it's a god.

CRAFT FAIR

Suppose you go to a craft fair and see a hundred booths of every kind of craft imaginable. The fair is held at a certain place and on a certain date—even certain hours of the day. Will you assume the fair came together like this accidentally? Did all of the people just happen to show up at the right time and place and date to attend this? Did the merchants just happen to be there? Did all of their fancy craft pieces just happen to get made and displayed? Is there order to the display? Did that order happen accidentally too? If a strong wind comes up during the day and tosses things around, will all of the displays get straightened out by themselves and end up in the right booths? How about if this wind storm happens over and over? Will everything eventually get back in the right place? We could even give it a hundred or a thousand or a million or billions of years. Will that be enough? It seems that something as organized as a craft fair takes some planning and would never happen accidentally, and if it ever got disorganized, it would take some more planning to get it back into a functioning unit. Then how could we figure that something as organized as a human body could happen without planning and organizing? There must be a planner, an organizer.

PHOTO LAB

Imagine if somebody really cruel went into a photo developing lab and dumped all of the photos, mixing up all of the envelopes. Let's say there were hundreds of pictures in envelopes, waiting to be picked up. Now all of the pictures are scattered. I wonder how long it would take before natural forces would organize those pictures again. Maybe some breezes would blow the pictures around. Maybe even a windstorm would happen on the scene. Without anyone directing the picture (pun intended), how long would it take to get all of the photos back into the correct envelopes?

Families would want to make sure that they got photos only of their own families, and they wouldn't want any of the pictures lost. If someone had pictures from an overseas trip, they'd want to get all of their own pictures back. It's not as if they could quickly redo the trip and try to capture the same scenes. Couples who had wedding pictures in there would want assurance that the right people were pictured in their packets.

I wonder how many years it would take to make all of this right again. Of course, it has to happen without any kind of organization, since there's no organizer. It has to all be a matter of time and chance. We will give it endless time and opportunity. All of the correct "families" will have to be together. All pictures of trips will have to have only pictures of that trip. That's what had to happen if we evolved, only on a much larger scale. Parts related to each other had to exist at the same time, and in the same place.

Speaking of families, have you ever had your picture taken for a family reunion? My family did several years ago, and there were 46 people in the picture, all descendants and spouses. I wonder how many years it would have taken for that picture to come together by itself. Given enough years, would all of the right colors land in the right spots on a piece of paper to look like everyone in my family? There's plenty of paint available, so that's no problem. There's enough paper. All we need is for the paint to be arranged correctly. What's so hard about that? If the people in the picture evolved out of nothing, it should be easy for a bunch of paint to evolve into only a *picture* of the same people. It must be harder for people to evolve than for a picture to evolve. I'm not asking the picture to come alive, like the people have. If I couldn't expect the picture to evolve, how could I expect the people pictured to evolve?

EXPLODING THE BIG BANG **81**

BIRDS AND AIRPLANES

I saw a picture of a bird flying under an airplane. We got our ideas about flight from watching birds. The wings of a plane dip to affect flight, as do a bird's wings. Would we say the plane and its wings evolved to do that—with no planning from people? Then how can we say a bird's wings evolved for flight, with no planning from anyone? The bird is far more complicated, since it is alive. It eats, grows, reproduces, feels climate changes, sees and hears other creatures. If a plane could do all that, it'd *be some plane*! It would especially be *some* plane if it happened without any information, any direction from a person.

JIGSAW PUZZLE

Shake up a 1000-piece puzzle in a bag and let them the pieces land wherever they want. Dump them over and over. Will they eventually form the picture on the box? Such nonsense! But that's evolutionary thinking. Given enough time, the picture should be complete.

DINNER

Someone invites us for dinner. We say it was terrific, but we tell the cook it came together by chance. The cook might suggest we come back tomorrow night and the cook won't be there. The meal will be whatever happens by chance.

MUSIC

We hear an orchestra play with all of the instruments in perfect harmony, and we know someone wrote the music and is coordinating the instruments. The musicians are all playing according to the same book. We know there is an organizer behind this.